W9-BRD-014

A Fair Bear Share

MathStart™ REGROUPING

by Stuart J. Murphy • illustrated by John Speirs

HarperCollins*Publishers*

LEVEL 2

To Maureen and Randy—
with hopes for their fair share of happiness
—S.J.M.

To Tilly and Siggy
—J.S.

The illustrations in this book were done in Pelican watercolors on Waterford cold press paper.

Bugs incorporated in the MathStart series design were painted by Jon Buller.

HarperCollins®, ■®, and MathStart™ are trademarks of HarperCollins Publishers Inc.
For more information about the MathStart series, please write to HarperCollins Children's Books,
10 East 53rd Street, New York, NY 10022, or visit our web site at http://www.harperchildrens.com.

A FAIR BEAR SHARE
Text copyright © 1998 by Stuart J. Murphy
Illustrations copyright © 1998 by The Big Cheese Design, Inc.
Printed in the U.S.A. All rights reserved.

Library of Congress Cataloging-in-Publication Data
Murphy, Stuart J., 1942–
     A fair bear share / by Stuart J. Murphy ; illustrated by John Speirs.
        p.   cm.   — (MathStart)
     "Level 2, regrouping."
     Summary: Four bear cubs collect ingredients for a blueberry pie, counting and
recounting them as their supply grows.
     ISBN 0-06-027438-7. — ISBN 0-06-027439-5 (lib. bdg.). —ISBN 0-06-446714-7
(pbk.)
     [1. Bears—Fiction.   2. Blueberries—Fiction.   3. Arithmetic—Fiction.
4. Counting.]   I. Speirs, John, ill.   II. Title.   III. Series.
PZ7.M9563Fai   1998   96-45026
[E]—dc21   CIP
                                                                              AC

Typography by Elynn Cohen
2   3   4   5   6   7   8   9   10
❖

A Fair Bear Share

Mama Bear said to
her four little cubs,

"If you'll collect enough nuts, berries, and seeds,
then tonight I'll bake my special Blue Ribbon Blueberry Pie.
And you'll each get a fair bear share."

The cubs went to the forest, where there were lots of sweet-tasting nuts. Three of them worked very hard.

But the fourth little cub just wanted to play.

She climbed trees.

She swung from branches.

She didn't look for even one nut.

The cubs ran home and gave their baskets to Mama Bear. She said, "If we put the nuts in groups of tens and put the ones that are left over next to the groups, then we can quickly add them up."

The cubs piled up the nuts they had collected by tens.

The first cub had 11.		"That's 1 group of ten and 1 left over."	11
The second cub had 14.		"That's 1 group of ten and 4 left over."	14
The third cub had 12.		"That's 1 group of ten and 2 left over."	12
The fourth cub didn't have any at all.		"That's 0 nuts."	+ 0

3 tens and 7 nuts

37

"That's 37 nuts altogether. Three of you did a very good job."

Next, the cubs went to the field to pick big, juicy blueberries.
Three of them picked really fast.

But the fourth little cub still wanted to play.

She skipped and ran.

She chased after bees.

She didn't pick even one berry.

11

The cubs ran home with their baskets again.
"Let's group the berries by tens and by ones
and add them all up," Mama Bear said.

The cubs piled up their berries by tens.

The first cub had 21.

"That's 2 groups of ten and 1 left over."

21

The second cub had 15.

"That makes 1 group of ten and 5 left over."

15

The third cub had 13.

"That's 1 group of ten and 3 left over."

13

The fourth cub didn't have any at all.

"That's 0 berries."

+ 0

4 tens and 9 berries

49

"That's 49 berries altogether.
Three of you did your share of the work."

Then the cubs went to the meadow to find some nice crunchy seeds.
Three of them grabbed all they could.

But the fourth little cub wanted to play some more.

She turned lots
of cartwheels.

She jumped all around.

She didn't grab even one seed.

15

Again they ran home and gave
Mama Bear their baskets.

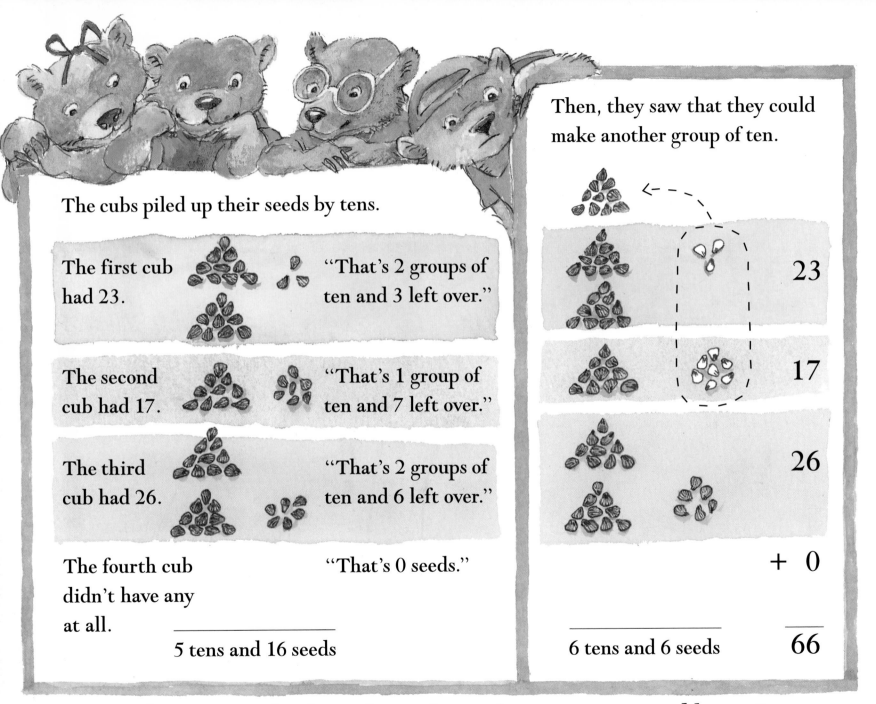

The cubs piled up their seeds by tens.

The first cub had 23. "That's 2 groups of ten and 3 left over."

The second cub had 17. "That's 1 group of ten and 7 left over."

The third cub had 26. "That's 2 groups of ten and 6 left over."

The fourth cub didn't have any at all. "That's 0 seeds."

5 tens and 16 seeds

Then, they saw that they could make another group of ten.

23
17
26
+ 0

6 tens and 6 seeds   66

"That's 66 seeds altogether. Three of you are very good bears."

Mama Bear looked at all the nuts, berries, and seeds that had been collected. She said, "I'm afraid there's not enough to make Blue Ribbon Blueberry Pie. Three of you worked very hard, I know, but one of you did not do a fair bear share."

The three hard-working cubs all looked at their little sister.

And she looked sadly down at the floor.

All of a sudden she jumped up,
took three baskets,
and hurried out the door.

First she ran to the forest
and found lots of nuts.

Next she skipped to the field and picked bunches of berries.

And she raced to the meadow and grabbed every seed she could.

Then she ran home, holding the baskets tightly in her little cub paws.

She proudly handed Mama Bear her baskets.
Mama Bear said, "Now it looks like you've
worked hard, too.

"Let's put things in piles by tens and by ones
and add them to all that we already have.

"We had 37 nuts.

That's 3 groups of ten and 7 left over."

The little cub had 15 more.

"That's 1 group of ten and 5 left over."

4 tens and 12 nuts

"We can make another group of ten."

37

+ 15

5 tens and 2 nuts   52

Mama Bear emptied
the next basket.

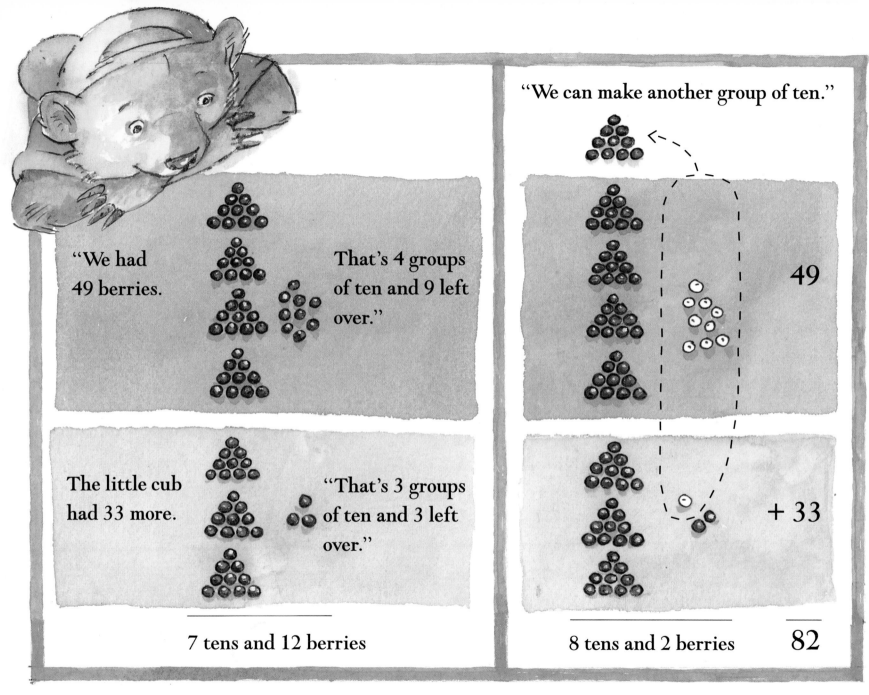

"We can make another group of ten."

"We had 49 berries."

That's 4 groups of ten and 9 left over."

49

The little cub had 33 more.

"That's 3 groups of ten and 3 left over."

+ 33

7 tens and 12 berries

8 tens and 2 berries   82

"All together, we have 82 berries."

Mama Bear emptied the third basket.

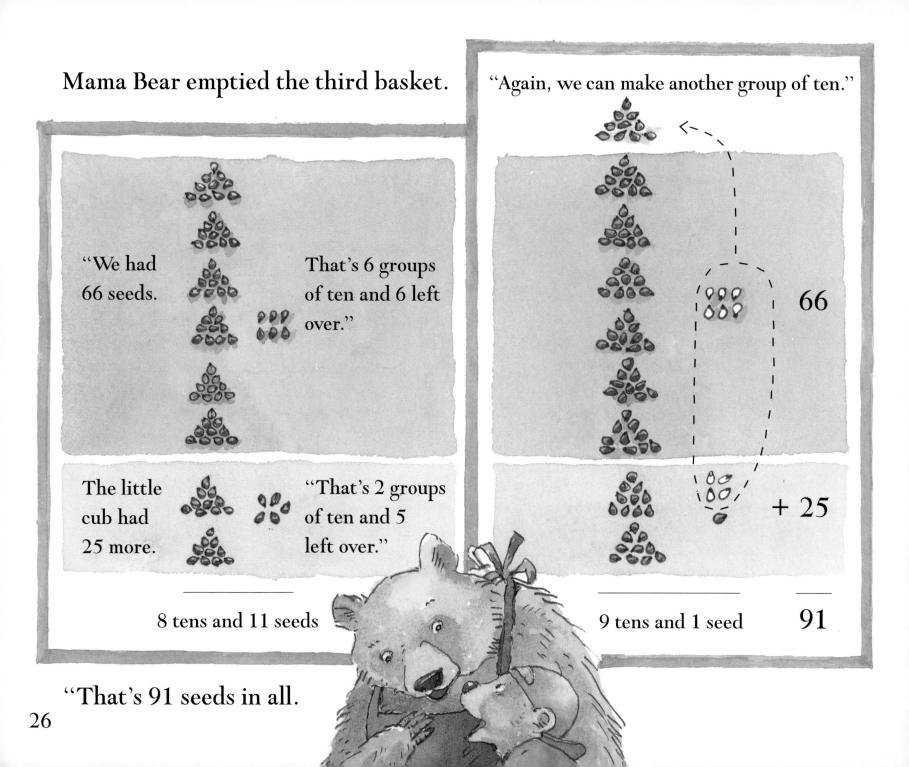

"We had 66 seeds.

That's 6 groups of ten and 6 left over."

The little cub had 25 more.

"That's 2 groups of ten and 5 left over."

8 tens and 11 seeds

"Again, we can make another group of ten."

66

+ 25

9 tens and 1 seed

91

"That's 91 seeds in all.

HOORAY!
Tonight I'll bake
Blue Ribbon
Blueberry Pie."

As they finished their dinner, the little cubs smelled
wonderful blueberry smells from the oven.

Mama Bear said, "I'm so proud of you all for working so hard.
The pie is almost ready, and each of you deserves a fair bear share."

Soon there was very little left—just a couple of nuts,

30

one or two berries, a few seeds, and four blue bear-cub smiles.

f you would like to have more fun with the math concepts presented in *A Fair Bear Share*, here are a few suggestions:

• Read the story with the child and describe what is going on in each picture. Ask questions throughout the story, such as "How many groups of ten can you make from 11? How many are left over?" Or, "How many groups of ten can be made from 49? How many are left over?"

• Encourage the child to retell the story. Use check marks to record the nuts, berries, and seeds that the cubs have collected. Circle the groups of ten.

• Look at things in the real world—crayons and markers, shells at the beach, stones or leaves at the park. Rearrange the objects by grouping them into tens and ones left over, and then add them all up. How many are there in all?

BOXES	CANS	BOTTLES
(✓✓✓✓✓) ✓✓ (✓✓✓✓✓)	(✓✓✓✓✓) ✓✓✓✓ (✓✓✓✓✓) ✓✓	✓✓✓✓ ✓✓✓

• Together, create an inventory of your kitchen cabinets. Count the boxes, cans, and bottles and make a mark for each. Circle the tens. How many items have you collected in all?

Following are some activities that will help you extend the concepts presented in *A Fair Bear Share* into a child's everyday life.

In the Car: Make a check mark on a piece of paper for each car, truck, or bicycle that you see. Group the marks into tens and ones. At the end of your drive add them all up. How many of each did you see? How many of all three did you see?

Taking a Walk: Walk around the neighborhood with some paper and a marker. Keep track of the number of kids, trees, dogs, or squirrels you see by marking down checks or dots. Count each category by circling the tens and then adding up the tens and the ones.

In the Playroom: Collect some crayons, markers, and colored pencils. Arrange each item into groups of tens and ones left over. How many are in each group? How many writing tools are in the playroom in all?

The following stories include similar concepts to those that are presented in *A Fair Bear Share*:

- ANNO'S COUNTING BOOK by Mitsumasa Anno

- COUNT! by Denise Fleming

- EACH ORANGE HAD 8 SLICES by Paul Giganti, Jr.